Sharks

Bull Sharks

by Julie Murray

Dash!
LEVELED READERS
1

An Imprint of Abdo Zoom • abdobooks.com

Dash! LEVELED READERS

Level 1 – Beginning
Short and simple sentences with familiar words or patterns for children who are beginning to understand how letters and sounds go together.

Level 2 – Emerging
Longer words and sentences with more complex language patterns for readers who are practicing common words and letter sounds.

Level 3 – Transitional
More developed language and vocabulary for readers who are becoming more independent.

THIS BOOK CONTAINS RECYCLED MATERIALS

abdobooks.com

Published by Abdo Zoom, a division of ABDO, PO Box 398166, Minneapolis, Minnesota 55439.

Printed in the United States of America, North Mankato, Minnesota.
102019
012020

Photo Credits: iStock, Minden Pictures, Shutterstock, © Doug Perrine / SeaPics.com p12, © David Kearnes / SeaPics.com p14, © David B. Fleetham / SeaPics.com p20
Production Contributors: Kenny Abdo, Jennie Forsberg, Grace Hansen, John Hansen
Design Contributors: Dorothy Toth, Neil Klinepier, Victoria Bates

Library of Congress Control Number: 2019941276

Publisher's Cataloging in Publication Data

Names: Murray, Julie, author.
Title: Bull sharks / by Julie Murray
Description: Minneapolis, Minnesota : Abdo Zoom, 2020 | Series: Sharks | Includes online resources and index.
Identifiers: ISBN 9781532129216 (lib. bdg.) | ISBN 9781098220198 (ebook) | ISBN 9781098220686 (Read-to-Me ebook)
Subjects: LCSH: Bull shark--Juvenile literature. | Sharks--Juvenile literature. | Fish--Juvenile literature. | Ocean animals--Juvenile literature. | Top predators--Juvenile literature. | Carnivores--Juvenile literature.
Classification: DDC 597.33--dc23

Table of Contents

Bull Sharks

What sharks can live in the salty ocean and a **freshwater** lake? Bull sharks!

Bull sharks live in **shallow**, warm waters. They are found in oceans around the world. They also live in rivers and lakes.

They have thick bodies. They are about 10 feet (3 m) long. They weigh 300 pounds (136 kg).

Bull sharks have small eyes. Their noses are short and round.

They are gray in color. Their bellies are white.

They eat fish, crabs, and shrimp. They also eat smaller sharks.

Bull sharks often swim slowly on the ocean floor. But they can move fast too. This makes them good hunters.

Bull sharks usually hunt alone. They are sometimes seen in groups.

Bull sharks can be **dangerous**. They can mistake humans for food and attack!

More Facts

- Bull sharks are one of the most **dangerous** sharks for humans to encounter.

- They are often seen in the Amazon River in South America.

- They have been found in the Mississippi River as far north as St. Louis.

Glossary

dangerous – likely to cause harm.

freshwater – having to do with water that is not salty.

shallow – not deep.

Index

Online Resources

Booklinks
NONFICTION NETWORK
FREE! ONLINE NONFICTION RESOURCES

To learn more about bull sharks, please visit **abdobooklinks.com** or scan this QR code. These links are routinely monitored and updated to provide the most current information available.